SpringerBriefs in Applied Sciences and Technology

More information about this series at http://www.springer.com/series/8884

Helga Kristjánsdóttir

Economics and Power-intensive Industries

 Springer

Helga Kristjánsdóttir
Bifröst University
Bifröst
Iceland

ISSN 2191-530X   ISSN 2191-5318   (electronic)
ISBN 978-3-319-12939-6   ISBN 978-3-319-12940-2   (eBook)
DOI 10.1007/978-3-319-12940-2

Library of Congress Control Number: 2014954591

Springer Cham Heidelberg New York Dordrecht London

Printed on acid-free paper

Springer is part of Springer Science+Business Media (www.springer.com)

Acknowledgment

This book is dedicated to my mother Olga, who was very supportive throughout the working process of the book. She taught me to tell a good SAGA, following the old Icelandic tradition of poetry and storytelling.

Contents

Chapter 1
Introduction

Where is the center of the universe? Some say it is gravitated somewhere between the trading centers of EU and the US, in the middle of the Atlantic Ocean. The entrance to the center of the earth is through Iceland according to a famous science fiction by Verne (1864).

The mid Atlantic Ocean location puts Iceland on the Europe–America flying routines, as well as on the potential sailing routines to Asia. An old Viking saga tells about Leifur Ericson, the son of Iceland and grandson of Norway, sailing to Iceland on his way to discover America. When considering sailing routines between continents during the Viking Age, Iceland's location was favorable, and it still is.

Being in the middle of the Atlantic Ocean (Fig. 1.1), Iceland is awash with abundant renewable energy resources, and halfway between Europe and the US. The country is characterized with highlands and glacial as well as volcanic landscape, and endowed with hydropower and geothermal energy. Iceland's natural richness of clean energy is of attention here, making it an interesting case of a renewable and sustainable energy driven country. Hydropower energy accounts for about three-fourths of all electricity production in Iceland, and geothermal energy for the remaining one-fourth. Due to its isolation the country is unable to export its abundance directly, and therefore does so indirectly through export of energy intensive products. This implies international trade and investment, since energy wholesale to the power intensive industry is required in order to take advantage of the abundant natural resource, and majority of the industry is owned by foreigners through foreign direct investment. This small open economy is interestingly positioned between the trade blocs of NAFTA and EU. In this current research, investment is explained by skilled labor and country size, among other factors. Green energy is without a question going to be a major factor in our future, so studying it at its nascence is particularly exciting.

In recent times, demand for renewable energy has been increasing substantially, following environmental awareness all around the world. This wave has reached the shores of the isolated island, Iceland, which has become a favorable chose of location for multinational companies in the power intensive industry. A crucial

© The Author(s) 2014
H. Kristjánsdóttir, *Economics and Power-intensive Industries*,
SpringerBriefs in Applied Sciences and Technology,
DOI 10.1007/978-3-319-12940-2_1

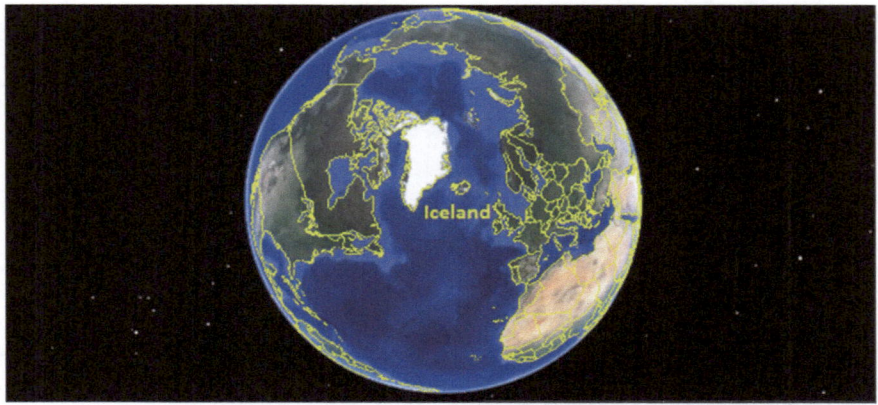

Fig. 1.1   Iceland in the middle of the Atlantic Ocean. *Source* Google Earth (2014)

element of this story are the driving forces for operations within the power intensive industry, applying Iceland's natural richness, in the form of hydropower and
geothermal power resources. Iceland is the highest electricity producer per capita in the world, and firms within the power intensive industry use majority of all
electricity produced in Iceland.

This book covers renewable and sustainable energy in several forms, under an
economic perspective. Readily available government and international statistics
are applied to provide insight on how businesses and economists can interpret factors influencing growth of the power intensive industry multinational firms like
Century Aluminum, Alcoa, Rio Tinto Alcan, Elkem and Becromal.

The journey then continues introducing modelling of all kinds, including usage
of the gravity model, with the gravity force being a permanent feature in earth's
evolvement. In the origin of earth and ever since, gravity has helped in creating
order and equilibrium. The basic specification of the gravity model combines distance and mass, together with other features. The knowledge capital model is also
accounted for, because of its allowance for incorporation of knowledge, measured
with secondary school education. These models have gained popularity in recent
economic literature.

The latter part of this saga relates to Eyjafjörður, fjord in the very north of
Iceland. In this fjord, some years ago, scientists found mineral rich smokers rising from the ocean floor. When the early earth was covered with water, mineral
rich smokers are believed to have played role in the creation of life. Under these
conditions, strokes from the features of smoking guns rose from bedrock to the
ocean, rich in dissolved minerals and gases like hydrogen and hydrogen sulphide.
Similar features are visible in Iceland, above sea level in the form of geysers. The
theme here is in part to examine the reversed version of the effects created from
smoking guns and geysers in the above saga, emphasizing the importance of earth
equilibrium to the benefit of nature and society, by injecting greenhouse gases like

carbon dioxide and hydrogen sulphide back to ground. In part this twilight saga implies contrast of the black smokers in an icy environment. In this sense the book presents a story of a time traveler, from the past to the present day. This involves introduction of how the current pollution is being turned around by injecting greenhouse gases into ground. This saga seeks to explain the development of the CarbFix and SulFix procedures in Iceland. These procedures involve injection of carbon dioxide CO_2 and hydrogen sulphide H_2S emissions back to the bedrock, reducing greenhouse gas emission when harnessing geothermal resources. Resulting in a condition closer to equilibrium, with less greenhouse gas effects, and a barometer on balance and earth stability.

Chapter 2
Microeconomic and Macroeconomic Issues in the Power Intensive Industry

Abstract Microeconomic and macroeconomic issues are highly important for the power intensive industry. There is a substantial fixed cost associated with starting operating an aluminum smelter, it is referred to as the threshold cost associated with investments in the industry. Investment of this kind is regarded as being a long-term investment, subject to microeconomic input factors involving the availability and cost of energy and labor. Macroeconomic factors include infrastructure, interest rates and inflation, as well as country membership to trade blocs like the EU, EFTA, or NAFTA affecting trading and investment opportunities.

Keywords Controlling stock · Economies of Scale · EEA · EFTA · Endowment · EU

Economic discussion is generally divided into micro economics and macro economics. Companies within the power intensive industry are highly subject to microeconomic and macroeconomic developments. To work on standard economic lines, their operations are dependent on micro-economic conditions, and the overall macro-economic environment is important for business operations in the global economy. Firms are involved in foreign direct investment, when having ownership of controlling stock in operations in a foreign country.

The micro-economic environment is the business operating environment that firms are faced with, when entering into operations. Financial contribution of the power intensive industry is dependent on aluminum prices, with the industry being sensitive to fluctuations in aluminum prices in the world market. Micro-economic factors in the operating environment include availability of electricity and water, as well as skilled labor. Productivity is dependent on the availability of these factors.

Micro-economic foundations of the gravity model were laid out by Bergstrand (1985) to incorporate micro-economic factors like prices into general economic modelling, when explaining the driving forces of exports. Products can vary between firms within industries, or they may vary between countries, referred to as product differentiation. Research by Bergstrand (1990) assumes product differentiation between firms rather than countries.

H. Kristjánsdóttir, *Economics and Power-intensive Industries*,
SpringerBriefs in Applied Sciences and Technology,
DOI 10.1007/978-3-319-12940-2_2

The nature of investments within the power intensive industry is such that they generally involve huge investments, involving high fixed costs in the beginning of operations. It therefore implies a cost increase in steps as operations expand. Step function application can therefore been suitable for capturing the fixed threshold cost, the Heckman's (1979) two-step procedure has for example proven useful for this type of research (Davies and Kristjánsdóttir 2010; Kristjánsdóttir 2012c).

Macro-economic issues include factors like government stability, and infrastructure. Also country membership to Regional Trade Agreements like the EU or NAFTA is of relevance. Iceland has EFTA membership, and is in the European Economic Area EEA. In a macro-economic perspective it is important to consider the export value for the economy and its contribution to gross domestic product.

Macro economic long-term conditions are important for firms within the power intensive industry, since investment within the industry is generally classified as being a long-term investment. Macro economic factors include electric energy supply, harbor access, and labor availability.

Macro-economic perspective is important when considering the primary production of aluminum in Iceland, with the majority of all aluminum ever made in world still being in use, due to recycling opportunities. Aluminum recycling has proven to be highly beneficial, since the recycling process only requires about 5 % of the energy required for the primary production.

Micro-economic environment accounts for input factors for production, like electric power. The electricity needed to produce 1 kg of aluminum is about 15 kW h. Also, 2 kg of aluminum oxide is required, derived from 4 kg of bauxite, and moreover 0.5 kg of carbon is needed for the production (Icelandic Association of Aluminum Producers 2014).

The main operating firms within the power intensive industry in Iceland include Century Aluminum, Alcoa, Rio Tinto Alcan, Elkem and Becromal.

Power intensive firm operation's have proven to be highly important to promote regional development by providing direct and indirect jobs for the local communities workforce and creating multiplier effects for firms in related industries. The contribution of these firms is substantial for the overall Icelandic economy.

2.1 Norðurál, Century Aluminum

Norðurál has operated an aluminum plant at Grundartangi since 1998. The plant is located in Hvalfjarðarsveit municipality in Vesturland region, Southwest of Iceland. The owner of the plant is Century Aluminum, headquartered in California USA. The operations were previously owned by Columbia Ventures Corporation until 2004. In 2013, the plant produced approximately 290,000 t of aluminum. The energy is obtained from the energy firms of Reykjavik Energy, Landsvirkjun the National Power Company of Iceland as well as HS orka, involving application of

Fig. 2.1 Century Aluminum in the south west of Iceland, at Grundartangi about 40 min drive from the capital city Reykjavík. *Source* Google Earth (2014)

hydroelectric and geothermal resources. The number of employees in 2013 was about 600 people.

In 2008, Norðurál started to construct a new aluminium smelter at Helguvík in the municipality of Reykjanesbær in Suðurnes region, Southwest of Iceland with the aim to produce 250,000 tons aluminum per annum. In 2013, the erection of the aluminium smelter came to a halt due to uncertainties regarding the energy delivery and disputes on the energy prices. That dispute has not been resolved as yet and it is unforeseen when production will start (Figs. 2.1, 2.2, 2.3 and 2.4).

2.2 Rio Tinto Alcan Iceland

The first aluminium smelter in Iceland started in 1969 in Straumsvík, Hafnarfjörður municipality, close to the capital city Reykjavík. It was originally operated under the name Icelandic Aluminum Company (ISAL) and owned by Alusuisse in Switzerland. The plant has been enlarged four times. The plant current owner is Rio Tinto, headquartered in London UK, bought the plant. Today it is operated under the name Rio Tinto Alcan. In 2013, about 300 employees worked

Fig. 2.2   Century Aluminum. *Source* Author's photo (2014)

Fig. 2.3   Century Aluminum. *Source* Author's photo (2014)

Fig. 2.4   Rio Tinto Alcan, Straumsvík in the neighbourhood of Reykjavík. *Source* Google Earth (2014)

for the company and the plant produced around 200,000 t of aluminum. The National Power Company of Iceland supplies all electricity to the plant (Figs. 2.5, 2.6, 2.7 and 2.8).

2.3   Alcoa Fjarðaál

ALCOA Fjarðaál started to operate an aluminum plant in 2007 and achieved its full operation in April 2008. The plant is located in Reyðarfjörður, Fjarðarbyggð municipality in the East region of Iceland. The plant has an annual production capacity of up to 350,000 t of primary aluminum. All electricity used by Alcoa is received from Landsvirkjun, the National Power Company of Iceland. There are about 480 employees working in the plant. Alcoa is a multinational firm, head-quartered in Pennsylvania in USA (Fig. 2.9 and 2.10).

2.4   Elkem

Elkem Iceland is as Norðurál located at Grundartangi in Hvalfjarðasveit munici-pality. The plant went into operation in 1979 and currently produces 120,000 tons of ferrosilicon per annum. It also produces refined metal with reduced content

Fig. 2.5 Rio Tinto Alcan in the background of the presidential residence. *Source* Author's photo (2014)

Fig. 2.6 Rio Tinto Alcan. *Source* Author's photo (2014)

Fig. 2.7   Rio Tinto Alcan. *Source* Author's photo (2014)

Fig. 2.8   ALCOA Fjarðaál at Reyðarfjörður in the east part of Iceland. *Source* Google Earth (2014)

Fig. 2.9   Alcoa Fjarðaál. *Source* Author's photo (2013)

Fig. 2.10   Alcoa Fjarðaál at Reyðarfjörður. *Source* Author's photo (2013)

of aluminum, carbon and titanium. Landsvirkjun, the national power company of Iceland, is the sole provider of electricity to Elkem (Figs. 2.11, 2.12, 2.13 and 2.14).

2.5 Becromal

Becromal Iceland started its operation in 2009 in Akureyri municipality, North East of Iceland. The company produces aluminum foil anodizing and receives electricity from Landsvirkjun, the National Power Company of Iceland (Figs. 2.15 and 2.16).

2.6 New Potential Firms

Other foreign companies have shown an interest in examining the potential in the field of investments in power intensive industry in Iceland. As in the projects mentioned above, the government has offered various incentives, including tax deductions by signing investment agreements with the following companies.

Fig. 2.11 Elkem Iceland is located at Grundartangi in Hvalfjordur in the western part of Iceland. *Source* Author's photo (2014)

Fig. 2.12   Elkem at Grundartangi. *Source* Author's photo (2014)

Fig. 2.13   Elkem. *Source* Author's photo (2014)

Fig. 2.14 Becromal is located at Krossanes close to Akureyri in the north. *Source* Google Earth (2014)

Fig. 2.15 Becromal factory in Akureyri, Iceland. *Source* Author's photo (2013)

Fig. 2.16 Becromal factory in Akureyri, Iceland. *Source* Author's photo (2013)

2.7 Silicor Materials

Silicor Materials plans to build and operate a plant in Grundartangi which will be designed for production of 160,000 tons of high-quality solar silicon and 30,000 aluminium by-products (master alloy) per annum. The project estimates to use approximately 90 megawatts (MW) and the power will be provided by Orka Náttúrunnar ohf and Landsvirkjun hf. It is envisaged that the plant will start production before end of 2016 and reach its full capacity in second quarter 2018. In the two year construction time of the plant, Silicor Material estimates that the number of employees will be approximately 400 people. Once the plant is in full operation a special workforce, with a number of highly educated personnel, will be employed. The total number of employees is estimated to 350 people. The current owner of the plant in Iceland is Silicor Materials Inc., headquartered in San Jose, California USA.

According to Terry Jester (2014), CEO of Silicor Materials Inc, the final decision to locate the plant in Iceland is subjected to low energy cost and good knowledge of the metals business in the country. She has also stated that the free trade agreement between China and Iceland signed in 2013, will have positively affect on decision of Silicor Materials.

2.8 PCC Silicon Metal Production Plant

PCC Bakki Silicon hf. plans to build and operate a silicor metal plant at Bakki in Húsavík, Norðurþing municipality in the north east of Iceland. The owner of the company is the German company PCC SE. The operation of the new plant is estimated to start in 2017 with up to 32,000 tons of production capacity. PCC will need about 58 MW of power for its production and has already signed a power purchase agreement with Landsvirkjun, the National Power Company. It is estimated that approximately 120 people will work at the plant when the operation commence and around 350 people over the construction period.

2.9 United Silicon

United Silicon plans to build and operate a metallurgical grade silicon production plant in Helguvík. The plant is designed for the production of 21,500 tons per annum. United Silicon estimates to use approximately 32 MW of electric energy from Landsvirkjun, the Icelandic National Power Company. It is envisaged that the Plant will start operating in mid-2016 and will reach its full capacity in 2017.

Chapter 3
The Power Intensive Industry, from the International Economic Perspective

Abstract International economic approach to the power intensive industry offers analysis under the perspective of geography and trade, a field that has gained increased attention in recent years. This involves incorporation of economics of scale by accounting for market size, captured with country population size and gross domestic product. A geographic dimension is also included by using geographical distance. Cultural distance measure developed by Hofstede, has also been applied. Moreover, trade cost and investment cost in various countries have proven a useful measure when determining investment development under the perspective of international economics.

Keywords Export ratio · Geographic dimension · Geography and Trade · Gravity Force · Hofstede Culture Index · Host country of FDI

The power intensive industry economic environment is becoming increasingly more international, with international economics being important, through the international economic activities in the Icelandic economy. In recent years the power intensive industry has accounted for a considerable share of exports from Iceland, and its contribution was particularly important following the 2008 financial crisis, when the country was more dependent on the power intensive industry and its spillover effects. Not only has the power intensive industry contributed significantly to the Icelandic economy through exports (Kristjánsdóttir 2012), but it has also played a major role in the foreign investment of the economy (Davies and Kristjánsdóttir 2010; Kristjánsdóttir 2010, 2013) (Fig. 3.1).

Over the course of time, the power intensive industry activities have expanded in Iceland, with geography and trade playing an important role. The importance of international factors like culture have also been incorporated, by use of the Hofstede culture index (Davies et al. 2008). This has resulted in more foreign direct investment within the industry, involving the operations of multinational firms (Kristjánsdóttir 2010) locating part of their operations in the country.

The power intensive industry in Iceland mainly consists of aluminum smelters owned by foreign firms. These are multinational firms diversifying their operations

© The Author(s) 2014
H. Kristjánsdóttir, *Economics and Power-intensive Industries*,
SpringerBriefs in Applied Sciences and Technology,
DOI 10.1007/978-3-319-12940-2_3

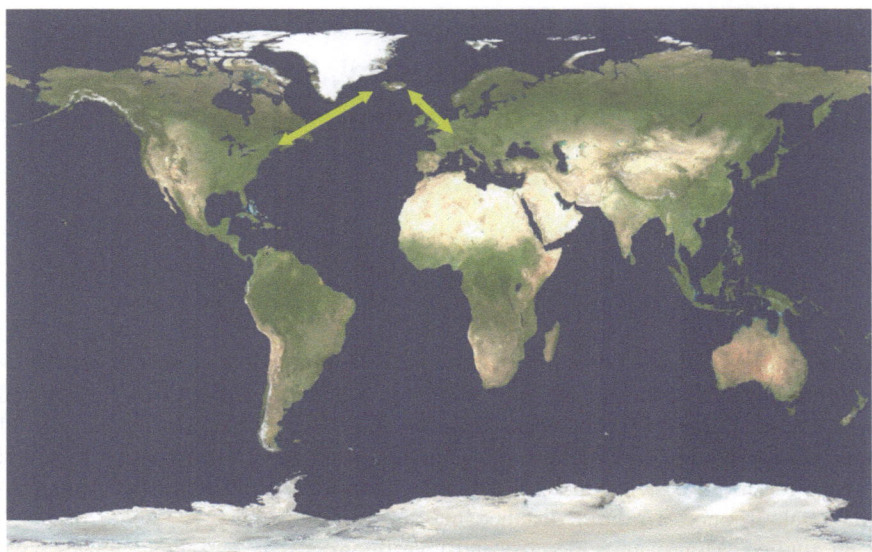

Fig. 3.1 Iceland main trade and FDI connections have been with the EU and NAFTA. *Source* Kristjánsdóttir (2004)

in different countries, dependent on the supply of the endowments needed for the production, including the endowment of electricity. The abundance of geothermal and hydropower energy has stimulated growth of firms within the power intensive industry, with their production being export driven. The operations of power intensive energy firms provide for opportunities of indirect energy export, through the export of power intensive products (Kristjánsdóttir 2012b). In 2013 the export share of aluminum in Iceland's exports by commodities was 35.3 %, and the share of Ferro-silicon was 3.3 % (Statistics Iceland 2014).

The raw material for used production, bauxite, is generally originated in the area close to the Equator, whereas the electricity needed for transformation into aluminum is obtained in places like Iceland. The international economic part of the story becomes visible when the bauxite is transported in tank ships from its origin to a country like Iceland, abundant with electric energy.

When considering the operating environment from an international economic perspective, transportation cost is highly important. Due to the location in the Mid-Atlantic Ocean, Iceland is well suited for sea transport between continents. This plays a part in choosing location for operating activities. Also, the geographic opportunity to locate activities in places with access to natural harbors in fjords can be of significant advantage, with bulk of the products needing tank ships for sea transportation. The geographical aspect has been incorporated into international economic theories, reflecting the industry operation effects on multinational entities (Markusen 2002). In essence multinationals base their activities on available resources, and the operating environment, as explained in the Knowledge capital model for multinational operations by Carr et al. (2001).

In recent years the Gravity Model and the Knowledge-Capital Model have gained popularity within the field of International Economics. The basic version of the Gravity Model implies that the dependent variable export is a function of gross domestic product among other things, reflecting the interaction of Keynesian variables. In a Knowledge-Capital Model the foreign direct investment (FDI) is a function of skilled labor among other things. Skilled labor is thereby accounted for as a measurable endowment, available in all countries to some degree. This provides the basic factors of the Cobb-Douglas production function, applied in the Solow growth model. Then market size is accounted for with population. In the spirit of Krugman it accounts for the new trade theory and the new economic geography. Incorporating economics of scale, increasing returns to scale, and imperfect competition as well as distance between countries. One can argue an inclusion of skilled labor as an endowment provides an opportunity for accountancy of the Factor Proportions Hypothesis (Heckscher-Ohlin Theorem) and thereby intra and inter-industry investment. When analyzing the importance of intra and inter-industry investment, it is common to account for investment cost and trade cost. These two cost factors are measured both for the source country, and the host country. The model sheds light on the conditions of horizontal and vertical investment. Similar size of countries and similar skilled labor are found to result in foreign direct investment in intra-industry, rather than inter-industry FDI.

Research by Kristjánsdóttir (2012a) analyses Iceland as a small open economy, alone in the middle of the Atlantic Ocean and highly dependent on trade with NAFTA and the EU. The research seeks to determine how important these trading blocs are to the country's exports, and how significant the country's isolation and small size is, and how these affect the export sectors. With the export volume typically being significantly impacted by the economic size of the exporting country, however it does not appear to apply to Iceland. This suggests that the exports from small remote economies are driven by different factors than exports from large economies. The data is analyzed using a unique transformation of the gravity model by an inverse hyperbolic sine function, allowing for accountancy of scattered exports.

Also a research by Kristjánsdóttir (2012c) studies power-intensive investment in a developed economy that boasts far more clean energy than the population can consume. The case country is Iceland, an isolated island that is unable to export its abundance directly and therefore must do so through foreign direct investment. Foreign direct investment is analyzed using a combination of the Knowledge-Capital and Gravity models. By combining these two models, both the effect of labor market skills and natural resources are taken into account when assessing foreign investments. Such a merger of data models can be used for other economies in a similar situation, who are interested in attracting investment in power intensive industries.

Furthermore, Kristjánsdóttir (2012b) investigates a new combination of the knowledge-capital and the gravity models. The model combination is applied to a small, remote country, which allows for testing a corner case solution. Furthermore, the substitution effects between inward and outward FDI are estimated by the use of a simultaneous equation system, and the estimates indicate that inward and outward FDI can be considered to be substitutes for each other.

Moreover, Kristjánsdóttir (2010) looks at how FDI in a small open economy in comparison to larger countries. Several specifications of the knowledge-capital model are applied to unique FDI data from the isolated country of Iceland, allowing for comparison with previous analysis of larger and similarly open economies. Using this, together with other techniques, the research seeks to explain investment determinants by geography, economic size and skilled labor availability. The results of these analysis show that popular specifications do not accurately predict the effects for a small country case.

Fixed costs play a crucial role in current models of foreign direct investment (FDI), yet they are almost entirely ignored in empirical treatments of FDI. This gap has been filled by using 1989–2001 panel data of FDI flows into Iceland, to examine the determinants of fixed costs for multinational firms, and how these influence aggregate patterns of investment (Davies and Kristjánsdóttir 2010). Additions to research in the field include usage of several natural resource variables, and the analysis of data on initial entry of FDI into a developed country. The Heckman two-step procedure is used, which allows for accountancy of fixed costs and their impact on estimation. Taken together, results indicate that application of the standard OLS approach to the data, tends to incorrectly link the quantity of FDI to source-country variables while in fact most of their role is in determining whether FDI takes place at all (Davies and Kristjánsdóttir 2010).

Iceland can be thought of as a small open economy, interestingly positioned between the trade blocs of NAFTA and the EU, with FDI in recent years resembling the pattern before the economic crash, making a pre-crash data set useful for exploring potential long-term trends. In a research by Kristjánsdóttir (2013), geographic dimension is accounted for, with investment being explained by geographic location, and country size. Gravity model is used to account for the country's exceptional remoteness and sparseness, a unique extension of the gravity model applies the inverse hyperbolic sine (IHS) function. The IHS functional form is estimated together with fixed difference between investment sectors and trade blocs being estimated simultaneously, analysis that is rarely possible. Results indicate that under these conditions, investment appears to be more driven by wealth than market size effects (Kristjánsdóttir 2013).

Paper by Davies et al. (2008) applies the panel fixed effects with vector decomposition estimator to three FDI data sets to estimate the impact of time-invariant variables on FDI while including fixed effects. Omission of fixed effects is found to significantly bias the results, leading to contradictory predictions regarding the effect of trade costs and culture across data sets. After eliminating these biases, the differences across data sets largely disappear and many time-invariant variables consistently indicate the importance of vertical FDI. This suggests that some controversies in the literature may be driven by the extent to which unaccounted fixed effects biases vary across different data sets.

Finally, some research has been performed to analyze the correlation between aid inflow and foreign direct investment inflow to the heavily indebted poor countries Malawi, Mozambique and Ghana. Data running from 1970 through 2004 are analyzed, using a simultaneous equation system to determine the interrelation

(Kristjánsdóttir 2012d). Due to the occasional small scale of flow, the inverse hyperbolic sine function is used, rather than a logarithmic function. Results indicate that when the sample countries experience a higher income per capita, complementary effects diminish at the cost of supplementary effects (Kristjánsdóttir 2012d).

Chapter 4
The Hydropower Application

Abstract One of the advantages of operating the power intensive industry is the potential access to clean energy, which is increasingly sought for with the demand for less environmental effects from operations. Hydropower energy is an example of clean energy, generated by harnessing waterfalls. Hydropower is commonly used within the power intensive industry for aluminum production. Majority of the hydropower generated in Iceland is applied by the power intensive industry.

Keywords Hydropower · Hydropower application

Hydropower is commonly applied source of power for intensive firms in Iceland, and most widely used form of renewable energy in the world.

Firms within the power intensive industry in Iceland use about 90 % of the electricity produced in Iceland (National Energy Authority 2014a). Hydropower energy accounts for about 75 % of the electricity production in the country (National Energy Authority 2014b).

Figure 4.1 shows Fljótsdalsstöð power plant, serving as the main electricity supplier for, ALCOA Fjarðaál.

Contracts regarding operation of multinational corporations within the power intensive industry have stimulated new hydropower construction, and been a driving factor for the hydropower related investments, driving up the energy use and making Iceland the highest electricity producer per capita in the world (Nation Master 2014) (Fig. 4.2).

Most of the hydropower plants in Iceland are owned by the National Power Company Landsvirkjun, making it the dominant electricity supplier in the country, and therefore the energy supplier for most of the companies in the power intensive industry in Iceland.

© The Author(s) 2014
H. Kristjánsdóttir, *Economics and Power-intensive Industries*,
SpringerBriefs in Applied Sciences and Technology,
DOI 10.1007/978-3-319-12940-2_4

Fig. 4.1 ALCOA Fjarðaál receives majority of its power from the Fljótsdalsstöð power plant, pictured here with the author standing inside. *Source* Author's photo (2013)

The hydropower production took off in 1969 with utilization of hydropower electricity production for the power intensive industry, serving aluminum smelter production in the south of Iceland.

Hydropower utilization advantage, over some other available renewable resources in Iceland, like the geothermal power, is that the hydropower provides for more stable energy supply into the energy system around the country. An electric energy is difficult to store after production, and therefore the hydropower production has advantages, since production can be managed dependent on demand. This is because the water can simply be stored in the reservoir and used when needed. The only possible shortcomings to energy provision in hydropower production is potential lack of water due to weather conditions in the highlands. In the case of hydropower, the gravity force is changing one form of energy into another form of energy.

Everything is energy, and the intensity and speed of one form of hydropower energy can be transformed to another type of electric energy. Hydropower energy is generated applying electricity generators to extract energy from falling or moving water, through the use of its gravitational force.

Fig. 4.2 Hot steam rising from a warm waterfall, created during making of a car tunnel through the mountain Vaðlaheiði in Eyjafjörður, Iceland. The waterfall reflects both hydro and geothermal resources. *Source* Author's photo (2014)

But what is gravity and why does it make sense to apply the gravity methodology under these circumstances? Let's look at the gravity concept in a wider setting, gravity is the force attracting objects to each other in proportion to their masses. It is common to talk about the gravity force of the planet, and how it affects the flow of physical masses, as well as wind and ocean currents. The law of gravity has been linked to the law of attraction. The gravity model in international economics is treating economic factors as the attracting or opposing forces, just as in physics (Figs. 4.3 and 4.4).

Fig. 4.3 a Cataclysmic events, such as this artist's rendition of a binary-star merger, are believed to create gravitational waves that cause ripples in space-time. *Source* NASA (2012a). **b** Flows around the globe with respect to gravity. *Source* NASA (2012b). **c** Volcanic eruption in Holuhraun in Iceland, during September 2014. *Source* TIME (2014)

Fig. 4.4   Power lines in Eyjafjörður. *Source* Author's photo (2013)

Fig. 4.5 The area around the waterfall Goðafoss on a cold September day. *Source* Author's photo (2014)

Fig. 4.6 Hydropower application is possible because of harnessing waterfalls. Waterfall in Bildsá and power lines in Kaupangur Eyjafjörður. *Source* Author's photo (2014)

Fig. 4.7   Holtavörðuheiði is part of the highlands in Iceland. *Source* Author's photo (2014)

Transport of hydropower and geothermal go together in Eyjafjörður Iceland (Figs. 4.5 and 4.6), with the highlands playing an important role in hydropower potentials (Fig. 4.7).

Chapter 5
The Geothermal Application

Abstract Power intensive industry operations involve clean energy applications including usage of geothermal energy. Application of geothermal energy has been successfully practiced in Iceland for multinational firms within the aluminum industry, and has accounted for about one fourth of all energy provision in the country. Firms within the power intensive industry in Iceland often use geothermal energy together with hydropower energy. Harnessing of geothermal resources involves minimal emission of geothermal gases, and current technology allows for reinjection of these gasses back to ground.

Keywords Geothermal application · Geothermal gas · Geothermal gas re-injection

One type of renewable energy is geothermal energy, and its growth has vast potential. Geothermal application by the power intensive industry, is mixed with hydropower application, making it less likely that the power intensive industry faces power shortage.

Geothermal sources play an important role in energy supply in Iceland, and have been harnessed for a long time, and their current application involve geothermal gas re-injection.

Some multinationals base their operations partly on the use of geothermal power, for example ALCOA and Century Aluminum. ALCOA operates in the east part of Iceland, using energy produced by the geothermal power plant Krafla together with energy supplied by Landsvirkjun, the National Power Company of Iceland, through Fljótsdalsstöð. Energy generated from harnessing of geothermal sources is supplied in various places around the country. The main power plants producing geothermal energy in Iceland include the power plants of Krafla, Hellisheiðarvirkjun, and Reykjanesvirkjun.

The geothermal sources that can be applied for the power intensive industries are those supplying steam, warm enough for turbines to generate electric energy.

© The Author(s) 2014
H. Kristjánsdóttir, *Economics and Power-intensive Industries*,
SpringerBriefs in Applied Sciences and Technology,
DOI 10.1007/978-3-319-12940-2_5

Fig. 5.1 The geothermal resources serving the power plant in Svartsengi are in the neighborhood of the Blue Lagoon in Iceland. *Source* Author's photo (2014)

Majority of geothermal energy production in Iceland is generated for the power intensive industry. Financing of geothermal projects tends to be more difficult than financing of hydropower projects, since these are considered to involve more risk (Kristjánsdóttir and Margeirsson 2012). To secure operations, geothermal companies like HS orka have long term contracts with power intensive firms like Century Aluminum, providing them with stable cash flow income. However, since the aluminum industry contracts tend to be subject to aluminum prices, it affects their capacity to pay for energy from suppliers (Figs. 5.1, 5.2, 5.3, 5.4, 5.5 and 5.6). Century Aluminum in Hvalfjörður is receiving energy from power plants in both Svartsengi and Reykjanes (Figs. 5.7, 5.8 and 5.9).

Fig. 5.2 Active geothermal resources create the foundation for the Blue Lagoon Iceland. *Source* Author's photo (2014)

Fig. 5.3 The power plant in Svartsengi behind the Blue Lagoon in Iceland. *Source* Author's photo (2014)

Fig. 5.4   The Blue Lagoon is derived from the Svartsengi power plant. *Source* Author's photo (2014)

Fig. 5.5   Vaðlaheiði, hot steam waterfall created in the making of a car tunnel in Vaðlaheiði close to Akureyri. *Source* Authors photo (2014)

Fig. 5.6 The warm ground by Lake Mývatn has been used since the beginning of settlement. For example there is a long tradition for using the ground's heat for baking bread, a process taking 22–24 hours. *Source* Author's photo (2013)

Fig. 5.7   Krafla power plant. *Source* Author's photo (2013)

Fig. 5.8   Krafla power plant. *Source* Author's photo (2013)

Fig. 5.9 Snowing in March the swimming pool in Iceland. *Source* Author's photo (2014)

The geothermal source is used for heating outdoor swimming pools, run both in summer and winter in Iceland.

Chapter 6
The Importance of Renewable Energy for the Power Intensive Industry, from an Economic Perspective

Abstract Economic research on the power intensive industry includes analyzing the importance of renewable energy for the industry. When doing so a sample for multinational activities is applied, often running over countries and years, and even industries. The Gravity model and Knowledge Capital models have proven useful for these analyses. The Gravity model is designed to capture volume of bilateral trade or investment. The Knowledge Capital model accounts for knowledge capital, measured by skilled labor. Also the Heckman 2 step procedure has proven useful, providing estimates in two steps to capture the threshold cost associated with investment in the industry.

Keywords Censored observations · Coefficient · Deflator · Dependent variable · Dummy vavriable · Dummy variable trap

The power intensive industry in Iceland is largely subject to renewable energy, and the greater awareness of environmental issues has made renewable energy a more feasible option. Hydropower availability has greatly added to the richness of the country, together with geothermal power, and made Iceland a favorable location for power intensive firms.

The renewable energy has become relatively more favorable when considering cost and pollution issues simultaneously, since pollution cost and quotas need to be taken into account when considering the overall cost and benefit analysis. Business environment mirrors the increased economic importance of using renewable resources in different countries. Increased expansion of the power intensive industry have to be carefully managed to preserve the environment (OECD 2104a).

The Third Kyoto Protocol on climate change implied that Iceland received a large pollution quota, compared to other countries engaged in the Kyoto Protocol, which may well have affected the sector allocation of FDI in Iceland (Kristjánsdóttir 2012c). The third Kyoto session was signed in 1997, for the Kyoto Protocol to the United Nations Framework Convention on Climate Change (1997). In the protocol Iceland has highest pollution quota of all the countries listed. An increase in pollution difference of the source and host country, with increase in

© The Author(s) 2014
H. Kristjánsdóttir, *Economics and Power-intensive Industries*,
SpringerBriefs in Applied Sciences and Technology,
DOI 10.1007/978-3-319-12940-2_6

the pollution quota of the host relative to the source, can be expected to increase investment in the power intensive industries in the host country, at the cost of the source country (Kristjánsdóttir 2012c).

Export of products from the power intensive industry are of major significance for the Icelandic economy, and have in recent years accounted for more than one third of the country's commodities export (Statistics Iceland 2014).

For a small economy like Iceland export is particularly important, since the economy is too small to be largely diversified in production, and therefore needs to depend on export to be able to import variety of goods and services (Kristjánsdóttir 2012b). Export gives an indication of the openness of countries.

Export ratio is the export of a particular country divided by its GDP, and the openness ratio is calculated as the sum of export and import divided by GDP. In general greater openness may effect economies so that they become more vulnerable to volatility due to trade shocks, however increased openness allows for specialization and scale economics (Kristjánsdóttir 2012b). The effects of the GDPs of the source country of export, and the recipient country of exports are sometimes accounted for when measuring international activities (Kristjánsdóttir 2012b, 2013). Also the interaction between the source country of FDI, and the recipient country has been measured (Kristjánsdóttir 2012a).

Theories on foreign direct investment assume certain threshold costs (Davies and Kristjánsdóttir 2010), these are threshold costs generally not dealt with in FDI empirical models (Davies and Kristjánsdóttir 2010). However, there is generally a certain fixed cost associated with investment, multinational entities (MNEs) need to consider when undertaking foreign direct investment (Markusen 2002). One way of accounting for the fixed costs associated with foreign direct investment is by applying the Heckman's (1979) two-step procedure reporting a Mills ratio to measure whether sample selection is driving the results (Davies and Kristjánsdóttir 2010, 2012c). Also, censored and uncensored observations can be applied when using the Tobit procedure (Kristjánsdóttir 2010) implying a subsample, a procedure proven useful when testing for fixed costs when entering into foreign direct investment (Fig. 6.1).

The Knowledge Capital model is often referred to as the KK model (Carr et al. 2001) incorporates fixed costs with its accountancy for investment costs and

Fig. 6.1 Foreign direct investment threshold cost. *Source* Kristjánsdóttir (2004)

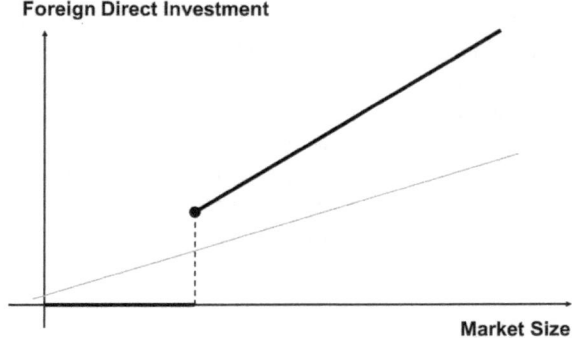

trade costs. When accounting for fixed costs, a combination of a gravity model and a Heckman two-step procedure has proven useful (Kristjánsdóttir 2012c). The Heckman procedure applies Probit estimates in the first step, accounting for whether investment is undertaken, and OLS coefficient estimates in the second step together with the R-squared measure to account for the level of investment. The procedure implies a selection equation, containing for at least one variable that is unrelated to the dependent variable used in the second stage, to prevent multicollinearity. Research has been made using the Heckman's (1979) two-step procedure to analyze one particular sector, the power intensive sector. The procedure has proven useful for accountancy of the substantial high fixed costs firms generally need to overcome, when undertaking investment within the power intensive industry (Kristjánsdóttir 2012c).

6.1 The Heckman Two-step Procedure

Figures 6.2 and 6.3 present the level estimates for FDI by the Heckman procedure for the gravity model.

When analyzing sector allocation of FDI in Iceland, Kristjánsdóttir (2012c) results turned out to be different from what was anticipated for the theoretical hypothesis of the KK model, since it does not perform very well for Iceland, although combined with the Heckman procedure. When estimating the regression equation results indicate (Kristjánsdóttir 2012c) that when the Heckman procedure

Gravity (LEVELS) Sample Selection

First Step Probit Results

Regressors	(1) Plain Vanilla	(2) Skill in St 1	(3) Skill in St 1&2	(4) Gvmt Stab.
ln(GMTST_OTHi,t)				-2.514**
				(-2.08)
ln(SKILL_OTHi)		3.326***	3.326***	4.678**
		(2.76)	(2.76)	(2.03)
ln(GDPt)	1.057	2.763	2.763	3.616
	(0.30)	(0.55)	(0.55)	(0.51)
ln(GDPi,t)	9.162***	17.728***	17.728***	28.870*
	(4.87)	(3.40)	(3.40)	(1.96)
ln(POPULATIONt)	-20.199*	-22.552	-22.552	-52.182
	(-1.73)	(-1.40)	(-1.40)	(-1.47)
ln(POPULATIONi,t)	-8.438***	-17.085***	-17.085***	-27.735**
	(-4.94)	(-3.53)	(-3.53)	(-2.02)
ln(DISi)	-1.485**	-1.560	-1.560	-3.361
	(-2.04)	(-1.05)	(-1.05)	(-0.93)
CONSTANT	-1.650	38.012**	38.012**	31.621
	(-0.17)	(2.26)	(2.26)	(1.09)

Note: Heckman's consistent Z - values are in parenthesis below coefficients. ***, ** and * denote significance levels of 1%, 5% and 10% respectively.

Fig. 6.2 Heckman Two-Step procedure results, First step. *Source* Kristjánsdóttir (2012c)

Gravity (LEVELS) Sample Selection

Second Step OLS Results

Regressors	(1)	(2)	(3)	(4)
	Plain Vanilla	Skill in St 1	Skill in St 1&2	Gvmt Stab.
ln(SKILL_OTHi)			1.183	0.628
			(1.64)	(0.75)
ln(GDPt)	6.525	7.353***	7.674***	7.309**
	(0.44)	(2.98)	(3.02)	(2.23)
ln(GDPi,t)	-9.577	9.905***	12.215***	9.424***
	(-0.28)	(4.80)	(5.06)	(5.48)
ln(POPULATIONt)	-7.816	-30.955***	-31.291***	-29.559***
	(-0.12)	(-4.33)	(-4.24)	(-4.08)
ln(POPULATIONi,t)	10.017	-8.184***	-10.583***	-7.967***
	(0.32)	(-4.34)	(-4.55)	(-4.78)
ln(DISi)	-3.529	-5.509***	-5.048***	-4.745***
	(-0.70)	(-8.65)	(-7.30)	(-4.26)
CONSTANT	-24.912	-5.794	7.467	0.318
	(-0.45)	(-0.68)	(0.59)	(0.02)
MILLS RATIO (λ)	-4.319	0.189	0.638	0.013
	(-0.58)	(0.33)	(1.12)	(0.03)
OBSERVATIONS	185	159	159	107
UNCENS. OBS.	36	36	36	27

Note: Heckman's consistent Z - values are in parenthesis below coefficients. ***, ** and * denote significance levels of 1%, 5% and 10% respectively.

Fig. 6.3 Heckman Two-Step procedure results, Second step. *Source* Kristjánsdóttir (2012c)

is combined with the gravity model, it appears to provide more significant results than when it is combined with the KK model, since the gravity model combination gives better indication of how host-country characteristics affect foreign direct investment (Kristjánsdóttir 2012c).

Figure 6.4 shows the interaction between various international economic factors, within the Edgeworth Box, in an easily understood visual way. In their research Carr et al. (2001) presented a graphical representation of the relationship between variables accounted for the KK model, as exhibited in the Edgeworth Box (Kristjánsdóttir 2010). A similar example of data accounting for variations in the host country's GDP is introduced in Markusen and Maskus (1999), with outward FDI from the US being the main source of data. An increased enlargement of the source country compared to host, corresponds to movement along the diagonal towards the NE corner in the Edgeworth Box. Indicating that along with increase in country size differences, FDI can be expected to decrease (Carr et al. 2001). This implies, that a small country like Iceland, generally viewed as a small economy and with skilled workforce, can be expected to be positioned in the northeast corner in the Edgeworth Box, since it is smaller than most source countries of investment, and has been considered to have skilled workforce (Kristjánsdóttir 2010).

In Fig. 6.5 the Inverse Hyperbolic Sine function and logarithm functions are exhibited. Gravity model specifications transformed with the inverse hyperbolic sine (IHS) function is provided by Kristjánsdóttir (2012b, 2013). The inverse hyperbolic sine function format can be more useful than the conventional logarithm

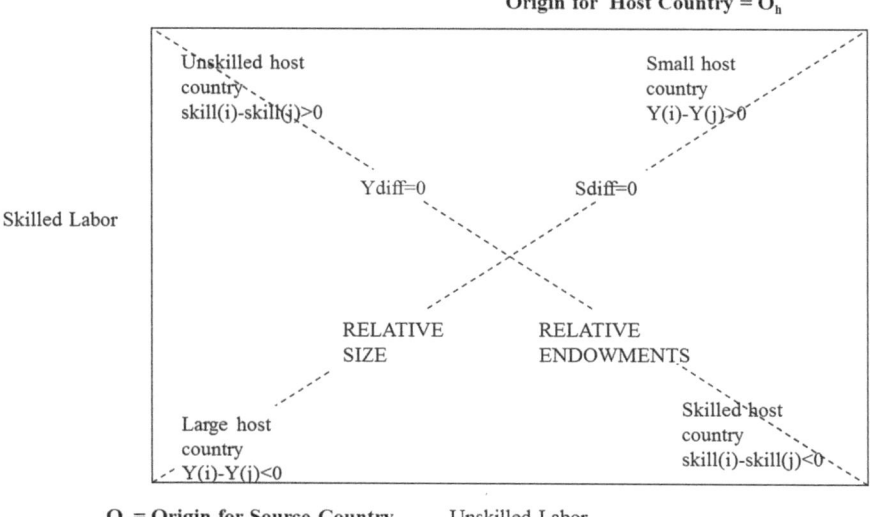

Fig. 6.4 The Edgeworth Box. *Source* Kristjánsdóttir (2010)

Fig. 6.5 The logarithm
and inverse hyperbolic
sine function. *Source*
Kristjánsdóttir (2012d)

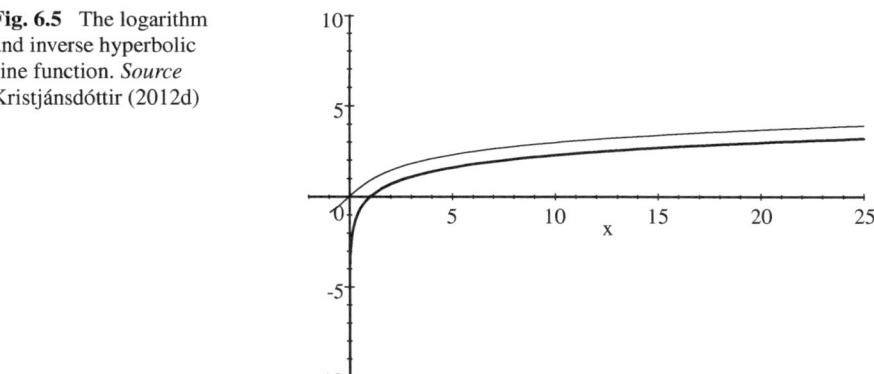

format, since the since inverse hyperbolic sine function allows for accountancy of zero and negative values. Such accountancy can be very useful when dealing with data on international activities in small economies like Iceland, since the FDI values in particular industries are not always positive (Kristjánsdóttir 2013), which in econometric terms indicates that it prevents unnecessary sample selection and potential non-normality of errors. The treatment of positive values, by the inverse hyperbolic sine function, provides a shape analogous to that of the logarithm function. This fact has proven convenient when investigating disaggregated data for countries, for example simultaneously for sectors and blocs (Kristjánsdóttir 2013).

In the international economic literature FDI is sometimes measured as affiliate sales (Brainard 1997), or as stocks or flows of FDI (Davies et al. 2008). With the FDI stock measure believed to provide a better indicator of the long-term strategies of multinationals (Davies 2008), since the stock measure carries information on previous investment incentives, until today. Some economists have preferred to use affiliate sales as an investment measure (Carr et al. 2001; Brainard 1997). However affiliate sales like FDI flows, tend to be subject to short-term, rather than long-term objectives of firms operations and therefore the FDI stock measure has proven useful since it reflects the long-term issues concerning international firms, including long-term factor price differences (Davies 2008). Factor price differences are important in factor proportions hypothesis, applied in international economics.

6.1.1 Data

The data applied in this research is data on foreign direct investment FDI in Iceland, based on the benchmark definition by OECD (2014b). OECD classifies FDI into the following five broad categories, (1) Primary Sector (Agriculture and Fishing, Mining and Quarrying), (2) Manufacturing, (3) Electricity, Gas and Water, (4) Construction, and (5) Total Services; as explained in the Benchmark Definition of Foreign Direct Investment, 3rd edition on the OECD webpage (OECD 2014b). The category of chose here is the second one, 2. Manufacturing, and within the category, the metal product industry is chosen (Fig. 6.6).

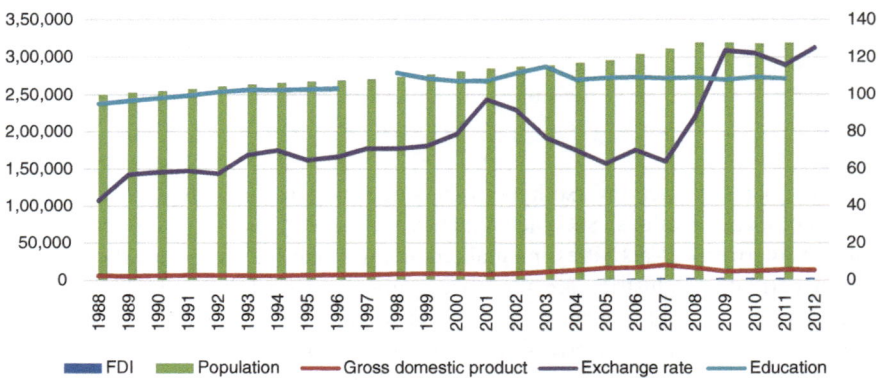

Fig. 6.6 Graphical presentation of the data used in this research. *Source* Author's computation based on data from the OECD (2014a) and the World Bank (2014)

6.1.2 Model Setup

The laws of nature are governing our everyday lives, and this accounts for the law of gravity. The universal law of gravity implies that the attraction of gravity weakens with the squared difference.

Equation (6.1) is based on the Bergstrand (1985) gravity equation and presents exports as a function of various variables, with $E(\ln u_{ij}) = 0$.

$$PX_{ij} = \beta_0 (Y_j)^{\beta_1} (Y_j)^{\beta_2} (D_{ij})^{\beta_3} (A_{ij})^{\beta_4} u_{ij} \tag{6.1}$$

The specification presented in Eq. (6.2) is a modified version of the Bergstrand (1985) specification of the gravity model:

$$FDI_{ij,t} = e^{\gamma_0} (Y_{i,t})^{\gamma_1} (Y_{j,t})^{\gamma_2} (D_{ij})^{\gamma_3} (A_{ij})^{\gamma_4} e^{\omega_{ij,t}} \tag{6.2}$$

The gross domestic product and population variables are presented in Eq. (6.3).

$$FDI_{ij,t} = e^{\gamma_0} (GDP_{i,t})^{\gamma_1} (POP_{i,t})^{\gamma_2} e^{\zeta_{ij,t}} \tag{6.3}$$

The equation is then set on a log-linear format, serving as a deflator by correcting for exponential growth and inflation in the economy. This logarithm of the explanatory variables equals the sum of their logarithms. The dependent variable is treated with an application of the inverse hyperbolic sine function which goes as follows $\sinh^{-1}(x) = \ln\left(x + (1 + x^2)^{0.5}\right)$. This linearization yields Eq. (6.4):

$$\sinh^{-1}(FDI_{ij,t}) = \varphi_0 + \varphi_1 \ln(GDP_{i,t}) + \varphi_2 \ln(POP_{i,t}) + \vartheta_{ij,t} \tag{6.4}$$

The modification then continues by undertaking further modifications of the model, as presented in Eq. (6.5):

$$\begin{aligned} \sinh^{-1}(FDI_{i,t}) = {} & \tau_0 + \tau_1 \ln(GDP_{i,t}) + \tau_2 \ln(POP_{i,t}) \\ & + \tau_3\, EXCH_{i,t} + \tau_4\, SKILLS_{i,t} + \kappa_{i,t} \end{aligned} \tag{6.5}$$

6.1.3 Variable Definition and Regression Results

The data runs over the annual period from 1988 through 2012, and is obtained from the OECD (2014a) and World Bank (2014). Values are reported in USD rather than ISK (Table 6.1).

The dependent variable $\mathrm{Sinh}^{-1}FDI_i, t$ accounts for inward FDI position in the metal product industry (OECD 2014a) in the host country of FDI, Iceland, on annual basis, reported in US dollar millions, and treated with the inverse hyperbolic sine function. The first explanatory variable $\ln GDP_{i,t}$ accounts for annual gross domestic product GDP is obtained from the OECD database (OECD 2014a)

Table 6.1 Variable definition

$Sinh^{-1}FDI_{i,t}$	Inward foreign direct investment (FDI) positions in the metal product industry, in the host country (i), over time (t) (OECD 2014a)
$lnGDP_{i,t}$	Gross domestic product (GDP), in the host country (i), running over time (t) (OECD 2014a)
$lnPOP_{i,t}$	Population, in host country (i) at time (t) (OECD 2014a)
$EXCH_{i,t}$	Exchange rate in the host country (i), over time (t) (OECD 2014a)
$EDU_{i,t}$	Skilled labor availability is proxied with education, secondary school enrollment, in host country (i) (World Bank 2014)

Table 6.2 Summary statistics for the basic sample

Variable	Units	Obs	Mean	Std. dev.	Min	Max
$Sinh^{-1}FDI_{i,t}$	US $, millions	25	6.623465	1.466834	4.450592	8.731891
$lnGDP_{i,t}$	US $, current prices, current exchange rates, millions	26	9.173508	0.3876436	8.628472	9.924673
$lnPOP_{i,t}$	Persons, all ages	24	12.54942	0.080075	12.42876	12.67406
$EXCH_{i,t}$	National currency per US dollar	26	79.48779	23.87024	43.014	125.083
$EDU_{i,t}$	Percentage of gross, secondary school enrollment	23	105.9311	5.178417	95.22467	115.0943

and reported in US $, current prices, current exchange rates, millions. The second dependent variable $lnPOP_{i,t}$ presents population, all ages, on annual bases (OECD 2014a). The third explanatory variable $EXCH_{i,t}$ accounts for exchange rates, period-average, measured in national currency per US dollar, as reported by (OECD 2014a). Finally, the forth explanatory variable education, accounts for school enrollment, secondary (% gross), and is obtained from the World Development Indicators (World Bank 2014) (Table 6.2).

In Table 6.3 the estimated based on Eq. (6.5), obtained for FDI in 1988–2012 are present.

The gross domestic product GDP is estimated to be significant. Some recent international economic literature has studied aspects of putting per-capita income back into trade theory (Markusen 2013). The positive estimate obtained for the GDP variable and the negative estimate for the population variable can be interpreted such the GDP per capita effects are positive. Also, the exchange rate EXCH is estimated to be significant, showing the effects of its development on foreign investors.

Table 6.3 Estimates for FDI 1988–2012

Regressors	
lnGDPi	3.699*** (8.01)
lnPOPi	−3.301 (−0.95)
EXCHij	0.0317*** (5.91)
EDUi	−0.0331 (−1.43)
Constant	15.213 (0.40)
R-sq	0.9813
Obs	23

Robust t-statistics reported in parentheses
***, **, * Significant at the 1, 5 and 10 % level respectively

All in all, the results in Table 6.3 indicate that FDI is significantly positively affected by host economic size and host exchange rate development over the period also has positive significant effects. The positive effects of exchange rate development is not surprising, since the host country currency weakened considerable during the economic crisis.

Chapter 7
The CarbFix Procedure

Abstract Scientists have sought for processes to reduce greenhouse gas emissions when harnessing geothermal resources, for the power intensive industry and other applications. Carbon dioxide CO_2 is among the greenhouse gases emitted when harnessing a geothermal reservoir. The CarbFix pilot program implies reduction of greenhouse gases by injecting and storing CO_2 in ground, a process referred to as carbon capture and storage. The procedure implies dissolving Carbon dioxide CO_2 in water and pumping it into basalt rock, to prevent it from entering into the atmosphere.

Keywords Bedrock · Capital costs · CarbFix · Carbon · Capture and storage · Carbon dioxide

Geysers and smoking guns are natural features originated in the early days of earth, emitting greenhouse gases like carbon dioxide (CO_2) and hydrogen sulphide (H_2S) to the atmosphere (Fig. 7.1).

Additional examples of greenhouse gases emissions related to geothermal activities, other than geysers and smoking guns, include emissions from geothermal utilization from both high- and low temperature fields. Recent technology has made it possible to reverse the emissions from a geothermal utilization back to the reservoirs, by mineralization. This circulation has provided an opportunity to bring about increased harmony in nature forces.

With more environmental awareness and understanding, countries have emphasized reduction of their CO_2 emissions, with the vision of bringing the universe closer to equilibrium in terms of reduction of greenhouse gases. These effects have been dealt with in Iceland when harnessing geothermal resources, emitting gases like CO_2, involving reinjection of greenhouse gases, through the CarbFix procedure (Ragnheiðardóttir et al. 2011), as well as the SulFix procedure (Gíslason and Oelkers 2014).

The driving forces for Reykjavik Energy entering into CarbFix was availability of CO_2 for experimentation, and basalt rock in the ground close to the power

Fig. 7.1 The area surrounding Geysir in Haukadalur in Iceland. *Source* Author's photo (2014)

plant, and the CarbFix procedure experience helped with developing the SulFix procedure (Júlíusson 2014). International agreements on emissions of greenhouse gases has motivated governments to regulation settings, and the Icelandic government has imposed regulations aiming at reducing emissions from power plants (Gunnarsson et al. 2013). Increased international obligations aim at reducing greenhouse gas emissions (Kyoto Protocol to the United Nations framework convention on climate change 1997) (Fig. 7.2).

CarbFix project objective was to develop a sustainable method for carbon capturing in basalt rock in order to reduce CO_2 emissions in air and thus reduce greenhouse gas submission (Gíslason and Oelkers 2014).

The CarbFix pilot program (CPP) implied reduction of greenhouse gases by injecting and storing CO_2 in ground, a process referred to as carbon capture and storage. The procedure implies dissolving CO_2 in water and pumping it into basalt rock. The CO_2 is dissolved in water, and injected down to the bedrock, where it gets attached to the iron in the bedrock by mineralization, a procedure presented in Eq. (7.1):

$$\left(Fe^{2+},\ Ca^{2+},\ Mg^{2+}\right) + CO_2 + H_2O = (Fe, Ca, Mg)CO_3 + 2H^+ \quad (7.1)$$

Equation (7.1) presents the chemical reactions when injecting CO_2 into ground, and it has been estimated that 8.8 tons of basaltic glass are required for each ton of

Fig. 7.2 The Krafla power plant in the north of Iceland. *Source* Author's photo (2013)

carbon to be fixed (Arnórsson 2003; Oelkers and Cole 2008; Oelkers et al. 2008; Gíslason et al. 2009).

Various carbon capture and storage procedures have been developed in other countries in Europe, for example for the German power market (Spiecker et al. 2014). However, regulation concerning greenhouse gas emissions varies between countries, and in the United States, the fragmented regulation has proven to be the primary barrier for the development to carbon capture and sequestration (Davies et al. 2013).

Cost factors associated with carbon capture and sequestration (CCS) projects have been estimated (Giovanni and Richards 2010), and the CCS project testing been developed in Iceland in recent years (Ragnheiðardóttir et al. 2011). Cost analysis for three scenarios of the CarbFix project, including the Hellisheiði full scale (HFS) scenarios by Reykjavik Energy, indicate that in the beginning the project is mainly subject to capital costs. However as the project increases in scale, cost factors dependent on electricity and water use become more significant. Calculations for return on investment involve net present value, IRR and EURIBOR (Ragnheiðardóttir et al. 2011) (Fig. 7.3).

The power company Reykjavik Energy lunched a project named CarbFix in 2007, in cooperation with scientists (Gíslason and Oelkers 2014). The CarFix project aimed at carbon capture and storage (CCS), a procedure undertaken through mineral carbonation at the Hellisheiði geothermal power plant (Aradóttir et al. 2011).

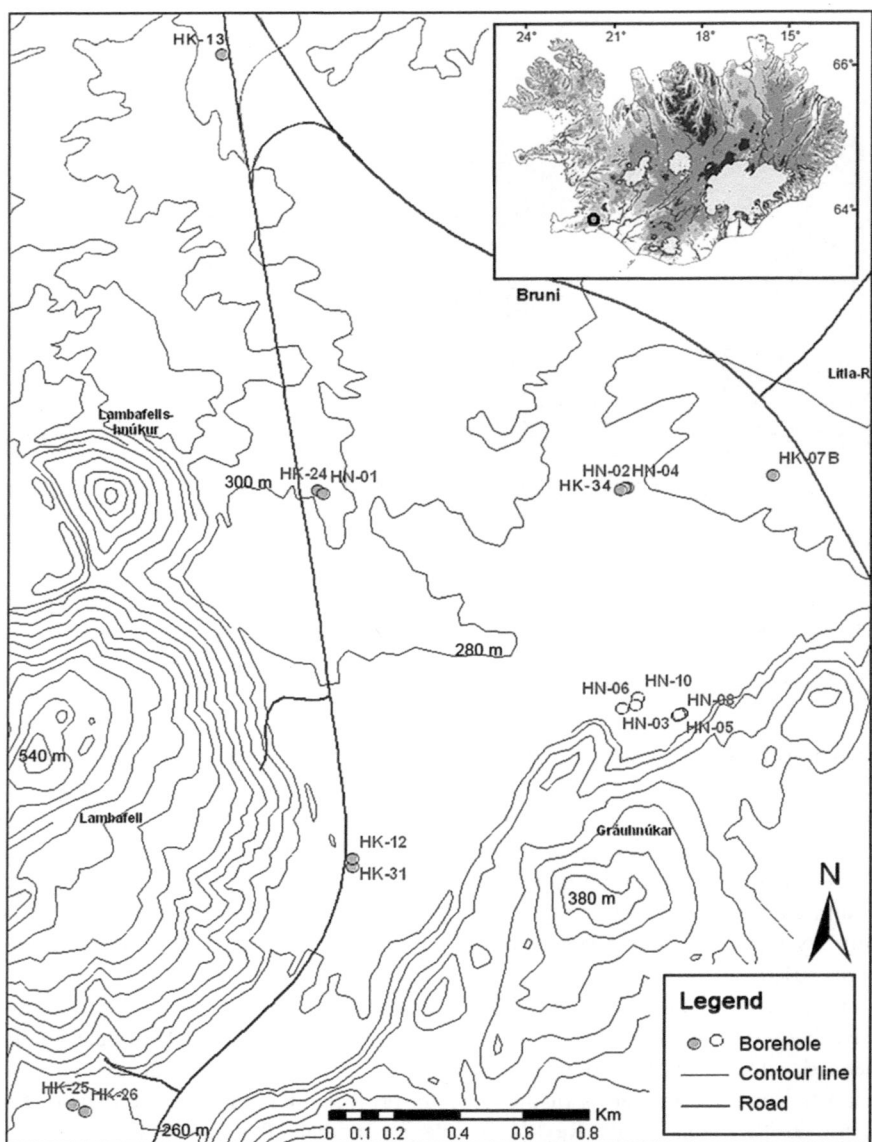

Fig. 7.3 The CarbFix project in Hellisheiði. *Source* Ragnheiðardóttir et al. (2011)

The power plant at Hellisheiði has proven to be a suitable basalt storage site, and the power plant submits a stream of non-condensable gases (Ragnheiðardóttir et al. 2011). Maximum reservoir exploitation is considered. Also costs associated with the CarbFix project at Hellisheiði have been evaluated and a pilot program has been scaled to higher flow rates of CO_2, with results indicating that in the beginning capital costs are most significant in the project overall costs, however as the CO_2 flow increases, the variable costs related to use of electricity and water become more relevant. Overall results indicate satisfactory returns on investment assuming 30 year life time, given possible trading prices at the time (Ragnheiðardóttir et al. 2011). In the range of 50–400,000 CO_2 tons the electricity is the highest variable cost factor, but once overstepping that threshold, the cost of water becomes more relevant.

$$Scaling\ Factor = \left[\frac{Flow_{scaled}}{Flow_{pilot}}\right]^{0.6} \tag{7.2}$$

Equation (7.2) presents the scaling factor applied in the CarbFix project, which is applied to figure out the increase in flow from the pilot program to the full scale program for energy requirements, and also to proxy equipment fixed cost increase (Ragnheiðardóttir et al. 2011) (Fig. 7.4).

The objective of the CarbFix project is to find a way to bind CO_2, the most common greenhouse gas, in a fixed-form deep in the bedrock the area surrounding

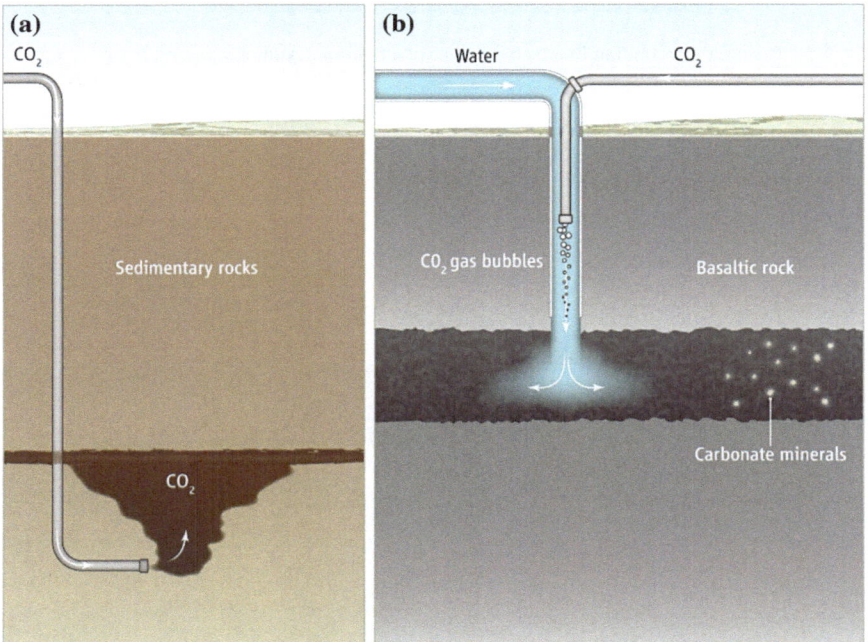

Fig. 7.4 The processes associated with carbon dioxide (CO_2) storage in basalt. **a** The process exhibited is the conventional process in capturing CO_2 into sedimentary rocks. **b** However, the process implies capturing of CO_2 dissolved in water, into basalt rock. *Source* Gíslason and Oelkers (2014)

Fig. 7.5 The area surrounding Svartsengi power plant. *Source* Author's photo (2014)

Fig. 7.6 The area surrounding Svartsengi power plant. *Source* Author's photo (2014)

the geothermal plant Hellisheiðarvirkjun run by Reykjavík Energy. This could reduce the CO_2 climate effects. The CO_2 is dissolved in water and injected into injection wells, and over time CO_2 binds to the rock and forms minerals (Gislason and Oelkers 2014).

More than 80 % of the CO_2 injected into ground has bound to basalt rock within a year (Gislason and Oelkers 2014) (Figs. 7.5 and 7.6).

Chapter 8
The SulFix Procedure

Abstract Application of geothermal resources for the power intensive industry, other industries and homes implies that greenhouse gases like Hydrogen Sulphide H_2S are emitted to the atmosphere. Hydrogen Sulphide is a gas that can be dissolved in water and injected into the ground, which is the objective of the SulFix project. Hydrogen Sulphide H_2S is dissolved in water and injected into the bedrock. The SulFix project is a unique project, conducted by Reykjavik Energy in Iceland, in the harnessing of geothermal resources.

Keywords Acid rain · Hydrogen sulphide · SulFix · Sulphur · Sulphur dioxide

Based on the progress of the CarFix experimental project, the SulFix project has been developed, applying similar methodology and procedures, expanding the procedures to industrial scale application (Júlíusson 2014). Scientists have been able to dissolve hydrogen sulphide in water and inject it into the bedrock (Júlíusson 2014). This type of procedure is the objective of the SulFix project undertaken at a power plant by Reykjavik Energy at Hellisheiði in Iceland, starting in industrial scale in the year 2014 (Fig. 8.1).

The harnessing of geothermal energy, implies emission of geothermal gases like hydrogen sulphide H_2S (Fig. 8.2).

The society calls for more green solutions regarding submission of greenhouse gases. International methods have been developed to deal with sulphide emission as by-product. These industrial processes generally involve different hydrogen sulphide H_2S abatement methods, with various of those being available and applied in different geothermal plants in the world, involving H_2S being converted to different forms through binding, burning or oxidation. This allows for storage of H_2S

© The Author(s) 2014
H. Kristjánsdóttir, *Economics and Power-intensive Industries*,
SpringerBriefs in Applied Sciences and Technology,
DOI 10.1007/978-3-319-12940-2_8

Fig. 8.1 Sulphur is visible in the neighborhood of the Blue Lagoon in Iceland. *Source* Author's photo (2014)

in sulphur, which for example can be seen piling up in hills in Alberta Canada and Vancouver USA, close to the borders of Canada (Júlíusson 2014). Hydrogen sulphide is also released through oxidization. Industrial activities involve H_2S being burned in the USA, transforming into SO_2 in the air, which can result in acid rain (Júlíusson 2014) (Fig. 8.3).

The most common way of releasing H_2S in geothermal industry, is releasing it to the atmosphere through cooling tower fans, where it slowly transforms into SO_2 in atmosphere, through oxidation (Júlíusson 2014) (Figs. 8.4 and 8.5).

The main driving force for undertaking the SulFix project was to reduce the H_2S concentration in the atmosphere, to improve air quality close to highly populated areas in the neighborhood of the power plants (Júlíusson 2014) (Figs. 8.6 and 8.7).

Local communities have been demanding more environmental solutions in harnessing geothermal resources, putting increased pressure on geothermal firms like Reykjavík Energy in reducing emissions from the power plants. The Hellisheiði power station is among the biggest single side geothermal power station in the world (Júlíusson 2014) (Fig. 8.8).

Three power companies Reykjavik Energy, Landsvirkjun the National Power Company of Iceland, and HS Orka, came up with techniques for reinjection of hydrogen sulphide into ground (Júlíusson et al. 2015). Conventional international abatement processes have involved by-products like sulphur and sulphuric acid,

Fig. 8.2 Geothermal steam in the air behind tourists enjoying the view at Námaskarð, close to Krafla in the North of Iceland, and the ground is colored by sulphur. In the past sulphur from that field was utilized as a mineral resource in Iceland, and exported for making gun powder. *Source* Author's photo (2013)

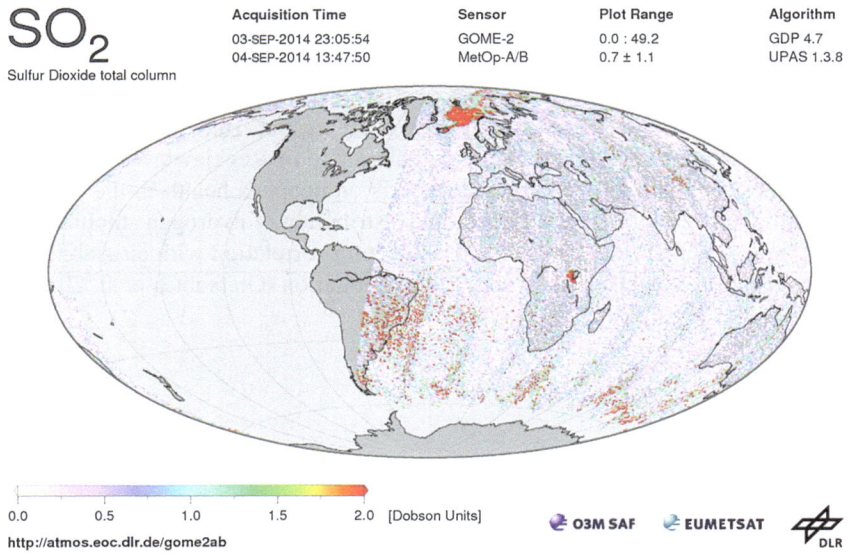

Fig. 8.3 The *red spot* in the *top* of the figure exhibits the distribution of Sulphur Dioxide SO_2 in the atmosphere, following volcanic eruptions in Iceland in September 2014. *Source* (O3M SAF 2014)

Fig. 8.4 The author standing next to the facilities on top of the reinjection boreholes, applied by Reykjavik Energy. *Source* Author's photo (2014)

and since there is small demand for these products in the Icelandic domestic market, they have not proven economically beneficial for geothermal firms in this market. The geothermal firms therefore decided to join forces and develop new processes for reinjection of gases into ground. These have been allocated and tested at industrial scale at Reykjavik Energy Hellisheiði Power Plant.

The economic vision of the SulFix venture is to provide a sustainable and environmentally friendly H_2S abatement procedure at low cost, making it more feasible than previously available commercial abatement options. The SulFix procedure involves mineralization of H_2S, by injecting it to ground, after dissolving it in water (Júlíusson et al. 2015) (Table 8.1).

Table 8.2 Estimated geothermal power plants in Iceland, CO_2 and hydrogen sulphide (H_2S) emissions.

Table 8.3 exhibits environmental limits for H_2S according to Icelandic regulation.

Ólafsdóttir et al. (2014) analyze spatial distribution of hydrogen sulfide resulting from utilization of geothermal resources at Hellisheiðarvirkjun and Nesjavallavirkjun power plants in the neighborhood of Reykjavik, with the objective to question the hydrogen sulphide effects on people's health in the neighboring communities. The spatial 30 km radius distribution of hydrogen sulphide in the neighborhood is therefore investigated, indicating correlation with air stability and low wind speed, as well as the absence of precipitation (Ólafsdóttir et al. 2014).

Fig. 8.5 Bjarni Már Júlíusson project manager inside the SulFix reinjection operating facilities. *Source* Author's photo (2014)

Fig. 8.6 Bjarni Már Júlíusson project manager at the SulFix injection site. *Source* Author's photo (2014)

Fig. 8.7 Bjarni Már Júlíusson project manager showing where gases are completely dissolved in water, at the SulFix site. *Source* Author's photo (2014)

Fig. 8.8 The gas separation station at Hellisheiði, for injecting hydrogen sulphide (H₂S) to the geothermal reservoir. *Source* Author's photo (2014)

Table 8.1 Greenhouse gas emissions from existing Icelandic geothermal power plants in operation

Icelandic geothermal power plants in operation	MW	CO_2	H_2S
Hellisheiðarvirkjun	303	44,934	12,374
Nesjavellir	120	14,794	8,709
Reykjanesvirkjun	100	25,090	860
Svartsengi	75	53,840	1,020
Krafla	60	39,683	5,180
Bjarnarflagstöð	3	1,292	1,603
Total emissions	661	179,633	29,746

Source Júlíusson et al. (2015)

Table 8.2 NCG emissions from Icelandic power plants in tons/year

Estimated emissions from planned geothermal power plants	MW	CO_2	H_2S
Hverahlíð	90		5,370
Reykjanesvirkjun	80	13,000	460
Eldvörp	50		550
Bjarnarflagstöð	45	2,980	2,744
Þeistareykir	90	19,690	9,845
Total emissions	355	35,670	18,969

Source Júlíusson et al. (2015)

Table 8.3 Environmental limits for H_2S according to Icelandic regulation

Environmental limits	Reference guidelines	Guidelines ($\mu g/m^3$)	Allowed repetitions over limits	Apply
Public health limits	Max average concentration for 24 h	50	5	From adoption of the regulation
Public health limits	Max average concentration for 24 h	50	0	July 1st 2014
Public health limits	Year	5		

Source Júlíusson et al. (2015)

Ólafsdóttir and Garðarsson (2013) research focuses on hydrogen sulphide concentration around the capital city of Iceland Reykjavik, from the emissions of the geothermal power plants in Nesjavellir and Hellisheiði in less than 35 km distance from the city. Findings indicate hydrogen sulphide can not only to be connected with emissions, but also related with the weather conditions in the area. Indicating that H_2S concentration can be predicted dependent on weather forecasts accounting for factors like wind and temperature (Ólafsdóttir and Garðarsson 2013).

Glossary

Acid rain Acid rain sometimes occurs when SO_2 is emitted to the atmosphere

Bedrock Ground below zero level, the bedrock in Iceland has proven particularly suitable for binding greenhouse gases, because how basaltic it is

Capital costs Capital Costs imply the fixed costs of setting up operations

CarbFix The CarbFix pilot program implies reduction of greenhouse gases by injecting and storing CO_2 in ground, a process referred to as carbon capture and storage. The procedure implies dissolving Carbon dioxide CO_2 in water and pumping it into basalt rock

Carbon capture and storage Carbon capture and storage (CCS) also referred to as carbon capture and sequestration, implies capturing CO_2 and storing it, to prevent it from entering into the atmosphere

Carbon dioxide Carbon dioxide CO_2 is greenhouse gas emitted when harnessing a geothermal reservoir

Censored observations Censored observations are those observations within a sample that are either above or below a certain value. Uncensored observations are those within the value range analyzed

Coefficient Coefficient estimate is a factor indicating how much the dependent variable changes when the explanatory variable changes by one unit. The coefficient sign obtained gives an indication of whether the change is positive or negative

Controlling stock Ownership of 10 % voting stock or more in a particular firm is sometimes referred to as controlling stock

Cost analysis Cost Analysis, also referred to as cost-benefit analysis, are analysis accounting for the risks and gains of projects

© The Author(s) 2014
H. Kristjánsdóttir, *Economics and Power-intensive Industries*,
SpringerBriefs in Applied Sciences and Technology,
DOI 10.1007/978-3-319-12940-2

CPP CarbFix pilot program

Deflator Deflator is an index used to take out the inflation, which is to deflate values. An example is the World Bank GDP deflator

Dependent variable The dependent variable is on the left side of the equation, and dependent on the change in other variables on the right hand side of the equation

Dummy variable An example of a dummy variable is a variable that takes a value of 0 or 1, dependent on if the country has a trade bloc membership or not. It can also take values of 1, 2, 3 and 4, dependent on which trade bloc the country belongs to

Dummy variable trap The case when the explanatory variables are multicollinear, when two or more of them are highly correlated

Economies of Scale (EOS) Economies of scale, also referred to as Increasing Returns to Scale is incorporated in the Gravity model by accounting for market size, proxied by country population size and gross domestic product (GDP). The incorporations of economies of scale allows for imperfect competition

Edgeworth Box The Edgeworth Box has been applied in International Economics when using the Knowledge-Capital model in explaining how size and skillness of host and source countries of investment, are related to foreign direct investment

EEA European Economic Area

EFTA European Free Trade Association

Empirical Specification Empirical specification is the specification of an empirical model

Endowment Endowment of an economy is the economy natural richness, abundance of natural resources

EU European Union

EURIBOR Euro Interbank Offered Rate

Exchange Rate Reflects the strength of one currency compared to another, it is the rate one currency is traded for another

Explanatory variable Explanatory variable is the independent variable on the right hand of the equation

Export The word export goods accounts for goods produced domestically and domestic services sold abroad, that exit a particular port, and export is most likely a shortening of the two words exit port

Export ratio Ratio of the export of a country to its gross domestic product (GDP)

Factor Proportions Hypothesis A popular approach when analyzing the determinants of Foreign Direct Investment (FDI) is to apply the factor proportions hypothesis, as to consider FDI dependence on factor endowments such as source and host country differences in skilled and unskilled labor. However, for small resource based economies like Iceland, the dependence on skilled and unskilled labor may not be the right endowment approach. Instead, resource based endowments need to be brought into the picture in order to reflect on the country's heavy dependence on marine and hydropower resources

Fixed cost Are the one time set up costs of a project

Geographic dimension Is sometimes included in the model by including distance

Geography and Trade There is a field of Geography and Trade within New Trade Theory, in which the Gravity Model is classified. The model incorporates economics of scale by accounting for market size, proxied by country population size and GDP. A geographic dimension is also included in the model by including distance

Geothermal application Refers to application of geothermal resources for energy provision of the power intensive industry

Geothermal gas In the harnessing of geothermal resources geothermal gases, like carbon dioxide (CO_2) and hydrogen sulphide (H_2S), are emitted to the atmosphere

Geothermal gas re-injection Geothermal gas re-injection in Iceland implies reinjection of greenhouse gases into basalt rock, as to minimize abatement from harnessing geothermal resources

Gravity Force The gravity concept is originated in physics, referring to Newton's law of gravity. Newton discovered the nature of gravity in his mother's garden in England, in the year of 1666, (Keesing 1998) when analyzing the pulling force causing an apple fall to the ground. He named this pulling force gravity. The gravitational force between two objects is dependent on their mass and the distance between them

Gravity Model When the gravity model is applied in international economics, exports or foreign direct investment correspond to the force of gravity, and gross domestic product corresponds to the economic mass. The gravity model is a macro model by its nature, since it is designed for capturing volume, rather than the composition, of bilateral trade

Gross Domestic Product The Gross Domestic Product (GDP) is generally applied in economics to measure the economic size of countries

Heckman two-step estimation The Heckman 2 step procedure estimates data in two steps. The first step implies a Porbit estimation, and the second step an OLS estimation

HFS Hellisheiði full scale

Hofstede Culture Index Cultural indicators developed by Hofstede, measuring differences in cultures between countries. The Hofstede culture index covers four dimensions of national culture: Power distance, individualism, masculinity, and uncertainty avoidance

Host country of FDI The country hosting the foreign direct investment (FDI)

Hydrogen sulphide Hydrogen Sulphide H_2S is a greenhouse gas emitted when harnessing a geothermal reservoir

Hydropower Hydropower is generated by harnessing waterfalls

Hydropower application Majority of the hydropower generated in Iceland is applied by the power intensive industry

Increasing Returns to Scale Returns to scale is also referred to as increasing returns to scale, and economies of scale. The term increasing returns to scale (IRS) is used when output increases relative to input, the term decreasing returns to scale (DRS) with relative increase in output, and constant returns to scale (CRS) when output proportion changes by the same proportion as input in production

Independent variable The variable explanatory variable on the right hand side of the equation. Opposed to the dependent variable, which is dependent on other variables

Inflation General price increase in the overall price level of an economy. Consumer price index (CPI) provides a measure for inflation

Infrastructural Factors Factors such as paved roads, internet and rail roads present infrastructure or infrastructural status

Injection well The well, or borehole, into which the greenhouse gases are injected to

International Economics Economic approach dealing with theories explaining trade, investment, movement of people, and financial transactions between countries, dependent on the operating conditions

Inverse Hyperbolic Sine function The Inverse Hyperbolic Sine function $\sinh^{-1}(x)$ is a function allowing for accountancy of zero and negative values. The natural logarithm function cannot operate on zero or negative values. The Inverse Hyperbolic Sine function procedure is therefore often preferred, because of the need for transformation that does not truncate or eliminate low values of the dependent variable. This way of imposing the Inverse Hyperbolic Sine Function (IHS) to the dependent variable while imposing natural logarithm on the independent variables has been used in studies on foreign direct investment, as well as on studies on household wealth

Investment Cost Investment cost is a measure provided to proxy the investment barriers facing investors entering the host country of investment

IRR Internal Rate of Return (IRR) is the discount rate at which NPV is equal to zero. Investors can view the IRR as a key figure by which alternative investment opportunities can be compared

ISK Icelandic Krona, in Icelandic called Íslensk króna (ISK) is the local currency of Iceland

KK model The knowledge capital model, is a model within the field of international economics accounting for knowledge capital, measured by skilled labor

Kyoto Protocol The Kyoto Protocol to the United Nations Framework Convention on Climate Change. The Third Kyoto session was signed in December 1997

Macro Economics Term providing an indication of the performance of the economy as a whole, as opposed to Micro Economics

Market Size Market size is sometimes proxied by population size of a country, in the performance of economic modelling

Metric tons Metric tons per year is a measure generally used in aluminum production

Micro Economics Term providing an indication of the performance of firms, individual industries or households within the economy, as opposed to Macro Economics

Mills ratio The Mills ratio implies whether the sample selection is driving the unconditional OLS estimates

Mineral carbonation The mineralization of CO_2 into basalt rock

Model Specification Model specification describes the relationship between the dependent and the explanatory variables

MRE Maximum Reservoir Exploitation

Multicollinearity Problems of multicollinearity arise when there is correlation between two or more of the independent variables

New Trade Theory The New Trade Theory allows for increasing returns to scale, imperfect competition, and product differentiation, covering both general equilibrium and partial equilibrium models of trade and trade policy

NPV Net Present Value considers the future cash flows when the time value of money is taken into account

OLS Ordinary Least Squares (OLS) is used for estimating linear regression model parameters. OLS is an estimation procedure that includes all observations regardless of their value

Openness Greater openness may cause economies to be vulnerable to volatility due to trade shocks, but more openness generally enables specialization and scale economics, resulting in faster economic growth

Openness ratio Openness ratio is the generally presented as the sum of export and import of a particular country, divided by its gross domestic product

Product differentiation Product differentiation refers to the difference in production of goods, between firms within industries or countries

Productivity Production process presenting output from one unit of input

Proxy Proxy variable is a measure used to approximate for another measure, for instance distance is sometimes inserted as a proxy for transport cost, and foreign affiliate sales to proxy FDI

Pseudo R squared Pseudo R squared indicates how the model fits the data, but is not an R squared in the general sense

R squared R squared is designed for linear models, and it is therefore not possible to calculate R squared for a non-linear model like the Tobit model. Therefore the so-called "Pseudo R squared" is calculated for the Tobit model

Recipient country of exports The country receiving exports, the importing country

Regional Trade Agreement Regional Trade Agreement (RTA) is a trade agreement between two or more countries, generally within a particular region

Regression Equation An equation listing the relationship between variables, with the dependent variable as function of one or more explanatory variables

Return on Investment Return on Investment (ROI) is the return gained on investment, dependent on interest rates

Sample Selection The Sample Selection procedure implies taking a subsample from a larger sample, using Mills ratios to imply if the sample selection is driving the unconditional OLS estimates

Scaling factor Scaling factor $= (\text{Flow-scaled}/\text{Flow-pilot})^{0.6}$

Sector Sector are composed of one or more industries

Significance Coefficient is said to be statistically significant if the probability is high enough that an effect is not due to just chance alone. Coefficient can be positive or negative, without being significant

Skilled Labor The measure for labor skillness is sometimes used to proxy skilled labor endowments. Skilled labor availability is sometimes proxied with secondary school enrollment

Source country of exports The exporting country

Source country of FDI The country undertaking foreign direct investment (FDI)

Subsample Subsample is a sample taken from a larger sample

SulFix The harnessing of geothermal energy implies emission of geothermal gases like hydrogen sulphide. This gas can be soluble in water an injected into the bedrock which is the objective of the SulFix project

Sulphur Sulphur is visible close to geysers and geothermal activities in Iceland. Hydrogen Sulphide can be stored in solid form as sulphur

Sulphur Dioxide Sulphur Dioxide SO_2 is created with oxidization of hydrogen sulphide, for example in industry related activities. Sulphur Dioxide is also released to the atmosphere during volcanic eruption

Theoretical hypothesis Implies theoretical guess about the relationship between factors

Threshold cost Threshold cost is the fixed cost investors sometimes need to overcome when making the initial investment

Tobit Tobit is an estimation procedure that accumulates all negative observations around zero. Thus, values lower than zero are set as zero and used as such for the regression estimates. Tobit estimates are consistent, if the error terms are normally distributed

Trade Bloc Trade bloc membership of countries involves trading agreement between them, with barriers to trade being reduced or eliminated between the member countries. Common trade blocs are NAFTA, EU and EFTA

Trade cost Cost associated with trading in a particular country. International economic studies often consider trade cost of the source country making the investment, and the trade cost of the host country receiving the investment

Trade Inter-industry Trade that occurs between industries

Trade Intra-industry Trade that occurs within a particular industry

Variable cost Company cost changing with output, opposed to fixed cost

References

Alfreðsson HA, Gíslason SR (2009) CarbFix—CO_2 sequestration in basaltic rock: chemistry of the rocks and waters at the injection site, Hellisheidi, SW-Iceland. Geochim Cosmochim Acta 73(13):A26–A26

Aradóttir ESP, Sigurðardóttir H; Sigfússon B, Gunnlaugsson E (2011) CarbFix: a CCS pilot project imitating and accelerating natural CO_2 sequestration. Greenhouse Gases-Science Technol 1(2):105–118

Arnórsson S (2003) Arsenic in surface—and up to 90 °C ground waters in a basalt area, N-Iceland: Processes controlling its mobility. Appl Geochem 18(9):1297–1312

Berden K, Bergstrand JH, van Etten E (2014) Governance and globalisation. World Econ 37(3):353–386

Bergstrand JH (1985) The gravity equation in international trade: some microeconomic foundations and empirical evidence. Rev Econ Stat 67:474–481

Brainard SL (1997) An empirical assessment of the proximity-concentration trade-off between multinational sales and trade. Amer Econ Rev 87:520–544

Carr DL, Markusen JR, Maskus KE (2001) Estimating the knowledge-capital model of the multinational enterprise. Amer Econ Rev 91:693–708

Century Aluminum (2014) Data. Available at: The Company. http://www.nordural.is/english/ obtained Mar 5

Davies RB (2008) Hunting high and low for vertical FDI. Rev Int Econ 16(2):250–267

Davies RB, Ionascu D, Kristjánsdóttir H (2008) Estimating the impact of time-invariant variables on FDI with fixed effects. Rev World Econ 144(3):381–407

Davies RB, Kristjánsdóttir H (2010) Fixed costs, foreign direct investment, and gravity with zeros. Rev Int Econ 18(1):47–62

Davies Lincoln L, Kirsten U, John R (2013) Understanding barriers to commercial-scale carbon capture and sequestration in the United States: an empirical assessment. Energ Policy 59:745–761

Emily G, Richards KR (2010) Determinants of the costs of carbon capture and sequestration for expanding electricity generation capacity. Energ Policy 38(10):6026–6035

Gíslason SR, Broecker WS, Oelkers EH et al (2009) The Carbfix project: Mineral CO_2 sequestration into basalt. Geochim Cosmochim Acta 73(13):A440–A440

Gíslason SR, Oelkers EH (2014) Geochemistry. Carbon storage in basalt. Science 344(6182):373–374

Gíslason SR, Wolff-Boenisch D, Stefánsson A et al (2010) Mineral seqestration of CO_2 in basalt—The CarbFix project. Geochim Cosmochim Acta 74(12):A336–A336

© The Author(s) 2014

H. Kristjánsdóttir, *Economics and Power-intensive Industries*,
SpringerBriefs in Applied Sciences and Technology,
DOI 10.1007/978-3-319-12940-2

Gíslason SR, Wolff-Boenisch D, Stefánsson A, Oelkers S, Gunnlaugsson E, Sigurðardóttir H, Sigfússon B, Broecker W, Matter J, Stute M, Axelsson G, Friðriksson Þ (2010) Mineral sequestration of carbon dioxide in basalt: a pre-injection overview of the CarbFix project. Int J Greenhouse Gas Control 4(3):537–545

Google Earth (2014) US Dept of State Geographer. Available at: Google Earth Downloaded Sept 7

Gunnarsson I, Aradóttir ES, Sigfússon B, Gunnlaugsson E, Júlíusson BM (2013) From Hellisheiði and Nesjavellir power plants, Iceland. GRC Trans 37:785–789

Jester T (2014) CEO of Silicor Materials. Available at: http://www.mbl.is/vidskipti/frettir/2014/08/21/island_heppilegt_fyrir_solarkisil/ obtained Aug 24.

Júlíusson BM (2014) Interview at Reykjavik Energy. Sept 11

Júlíusson BM, Gunnarsson I, Matthíasdóttir KV, Markússon SH, Bjarnason B, Sveinsson OG, Gíslason T, Thorsteinsson HH (2015) Tackling the Challenge of H_2S Emissions, Proceedings World Geothermal Congress 2015, Melbourne, Australia, April 19–25

Kristjánsdóttir H (2004) Determinants of exports and foreign direct investment in a small open economy. PhD presentation at the Univeristy of Iceland

Kristjánsdóttir H (2010) Foreign direct investment: The knowledge-capital model and a small country case. Scott J Polit Econ 7(5):591–614

Kristjánsdóttir H (2012a) Substitution between inward and Outward foreign direct investment. Publ Municipal Finance 1(2):23–28

Kristjánsdóttir H (2012b) Exports from a remote developed region: analyzed by an inverse hyperbolic sine transformation of the gravity model. World Econ 35(7):953–966

Kristjánsdóttir H (2012c) Knowledge is power: knowledge-capital model in the management of power intensive industries. Int J Energ Sect Manag 6(1):91–119

Kristjánsdóttir H (2012d) Talking trade or talking aid? Does investment substitute for aid in the developing countries? Open J Econ Res 2(2):3–20

Kristjánsdóttir H (2013) Foreign direct investment in a small open economy. Appl Econ Lett Taylor & Francis 20(15):1423–1425

Kristjánsdóttir H, Margeirsson Á (2012) Geothermal cost and investment factors. In Sayigh A (ed) Comprehensive renewable energy, vol 7. Elsevier, Oxford, pp 259–270

Kyoto Protocal to the United Nations framework convention on climate change (1997) Third session Kyoto, 1–10 December. Obtained from www.cnn.com/SPECIALS/1997/global.warming/stories/treaty/index4.html downloaded May 10th, 2002

Lonely Planet (2013) Available at: http://www.lonelyplanet.com/europe/travel-tips-and-articles/77734 obtained Mar 5

Markusen JR (2002) Multinational firms and the theory of international trade. MIT Press, Cambridge

Markusen JR (2013) Putting per-capita income back into trade theory. J Int Econ 90(2):255–265

Meuser W (1981) Obliterating old radical cavities by means of sulfix with reconstruction of the sound pressure transfer mechanism. Archives of Oto-Rhino-Laryngology-Archiv Fur Ohren-Nasen-Und Kehlkopfheilkunde 231(2–3):594–596

Ministry for Foreign Affairs (2014) Free trade agreement between the government of Iceland and the government of the people's republic of China. Available at: http://www.mfa.is/media/fta-kina/Iceland-China.pdf obtained Aug 24

NASA (2012a) Available at http://www.nasa.gov/topics/technology/features/atom-optics.html Retrieved 6.12.12

NASA (2012b) Available at http://heasarc.gsfc.nasa.gov/docs/objects/heapow/archive/active_galaxies/mcg_6_30_15_xmm.html Retrieved 6.12.12

Nation Master (2014) kWh per capita. Available at: http://www.nationmaster.com/country-info/stats/Energy/Electricity/Production/Per-capita obtained Sept 18

National Energy Authority (2013) Gas emissions from geothermal power plants. Available at: http://www.nea.is/the-national-energy-authority/energy-statistics/gas-emissions-from-geothermal/ obtained Mar 5

National Energy Authority (2014a) Power intensive industries. Available at: www.nea.is/hydro-power/power-intensive-industries obtained Mar 6

National Energy Authority (2014b) Energy statistics. Generation of electricity in Iceland. http://www.nea.is/the-national-energy-authority/energy-statistics/generation-of-electricity/ obtained Sept 18

O3M SAF (2014) The Satellite Application Facility on Ozone Monitoring (O3M SAF). Finnish Meteorological Institute (FMI). Available at: http://o3msaf.fmi.fi/pics/nrt/LATEST.METOPAB.GOME2.SO2.PGL.jpeg obtained Sept 4

OECD (2014a) Statistics. Available at: http://www.oecd.org/statistics/ obtained Sept 7

OECD (2104b) OECD Benchmark Definition of Foreign Direct Investment—4th edn. Available at: http://www.oecd.org/fr/daf/inv/statistiquesetanalysesdelinvestissement/fdibenchmarkdefinition.htm obtained Aug 24

Oelkers EH, Cole DR (2008) Carbon dioxide sequestration: A solution to a global problem. Elements 4:305–310

Oelkers EH, Gíslason SR, Matter J (2008) Mineral carbonation of CO_2. Elements 4:333–357

Ólafsdóttir S, Garðarsson SM (2013) Impacts of meteorological factors on hydrogen sulfide concentration downwind of geothermal power plants. Atmos Environ 77:185–192

Ólafsdóttir S, Garðarsson SM, Andradóttir HO (2014) Spatial distribution of hydrogen sulfide from two geothermal power plants in complex terrain. Atmos Environ 82:60–70

Ragnheiðardóttir E, Sigurðardóttir H, Kristjánsdóttir H, Harveyd W (2011) Opportunities and challenges for CarbFix: an evaluation of capacities and costs for the pilot scale mineralization sequestration project at Hellisheidi, Iceland and beyond. Int J Greenhouse Gas Control 5(4):1065–1072

Rio Tinto Alcan (2014) Available at: http://www.riotintoalcan.is obtained Mar 6

Spiecker S, Eickholt V, Weber C (2014) The impact of carbon capture and storage on a decarbonized German power market. Energ Econ 43:166–177

Statistics Iceland (2014) Exports by commodities (SI classification) 1999-2013. Available at: http://www.statice.is/?PageID=1261&src=https://rannsokn.hagstofa.is/pxen/Dialog/varval.asp?ma=UTA02105%26ti=Exports+by+commodities+%28SI+classification%29+1999-2013%26path=/Database/utanrikisverslun/UtflutningurAR/%26lang=1%26units=Fob%20value%20million%20ISK/tonnes obtained Sept 18

TIME (2014) Look at these incredible close-ups of a Volcanic Eruption in Iceland. World Iceland. Obtained from: http://time.com/3248111/bardarbunga-volcano-photos-close-up-iceland/ downloaded Sept 4

Verne J (1864) Journey to the centre of the Earth. France, Paris